FOUNDATION

FOUNDATION PRESCHOOL
KINDERGARTEN BASE NUMBER

BY

MS. C

1stBooks - rev. 01/30/04

ACKNOWLEDGEMENT

THANKS TO ALL: FIRST GOD.
My beloved, Mother Ms. Mary L. Credit;
George Deer Thank You;
ICON printing; The Libraries; The Mathematic Professors;
All of our Critics;
All of our Supporters
The Buyers of this Book, and;
The Publisher, 1st BOOKS' LIBRARY.

Dedicated To Each Child.

Clara D. Credit, Author/ Composer. Education: Harding University, Bible, Searcy Arkansaw; International Bible College, Bible, Florence Alabama; Domestic Health Care Institute, Medical Administrative Assistance, Baton Rouge Louisiana; Bob Brooks School of Insurance and Realtor, Insurance, Baton Rouge Louisiana; Delta Junior College, Office Spectalist, Baton Rouge Louisiana; ALCORN STATE UNIVERSITY, Office Administrative, Lorman Mississippi; Wilkinson County Elementary and High School, Woodville Mississippi.

HELL**O** EVERYBODY,

My Name Is **AMICABLE,** and I am Your FOUNDATION PRE-SCHOOL KINDERGARTEN BASE NUMBER Counting Book Animated Tour guide. So you will be hearing a lot of my WORDS out of my Mouth. Although I am not seen yet! I AM your PEACEFUL, NEIGHBORLY Voice FRIEND, and I Actually Exist. I must Remain Invisible in this book, FOUNDATION PRE-SCHOOL KINDERGARTEN BASE NUMBER COUNT so that YOU can GET to KNOW ME; FIRST by my PERSONALITY. You see I am FRIENDLY, TOUCHABLE, LOVABLE, and VERY, VERY, VERY HUGGABLE. I am GOODWILL, FRIENDLY, NEIGHBORLY, and PEACEABLE this is what my Name AMICABLE mean.

1

I SHARE Meaning Element with the Characters In this Book. FOUNDATION PRESCHOOK KINDERGARTEN BASE NUMBER COUNT: <u>I</u> too <u>AM</u> exhibiting <u>GOODWILL</u> and <u>AN</u> <u>ABSENCE</u> of <u>ANTAGONISM</u>.

So just RELAX, as <u>I tour,</u> <u>READ,</u> and COUNT With You in:

THE FOUNDATION PRESCHOOL KINDERGARTEN BASE NUMBER COUNT, COUNTING, READING, WRITING, and ALPHABET book.

The Perfect Number Counting Book.

COME WITH ME, IN THE LANDS OF WONDERFUL; ON THE ROAD, IN THE PERFECT NUMBER COUNTING.

In the **BEGINNING, 0.** Beginners Engineers. PRE-SCHOOLERS KINDERGARTEN and ELEMENTARY SCHOOLS ENGINEERS are MATHEMATIC ACCURATE Incline USERS.

An Accurate Mathematic FOUNDATION for intended, (s long lasting, Safe function ROADS Streets), BUILDING and PLAYGROUNDS A MUST to hold up the building Build.

ONCE UPON A TIME A LONG LONG TIME AGO IN THE LAND OF THE ROMAN THE ROMAN NUMBERS STAND IN GREAT POWER.

ROME HAD CONQUERED A NUMBER OF TERRITORY; ESTABLISHING THE ROMAN EMPIRE. THIS GREAT MONACH THRONE RULED IN EXTREMELY GREAT POWER. WITH ROME AS THE GREAT MONACH RULING ON THE THRONE, ITS' SUBJECTS, THE ROMAN NUMBERS RULED ALSO BEING ON THRONE WITH IT EMPIRE.

HISTORICAL HISTORY, the ROMAN NUMBER Were the most World Widely used Numbers in Schools, for EDUCATION Counting and Writing. Therefore All of SOCIETY everywhere, Counted and Wrote In THE ROMAN NUMBER COUNT and WRITE all the NATION. (Rome/Hebrew).

THEN
The UpRise of the ARABIC NUMBER
Writing and COUNTING 0,1,2,3,4,5,6,7,8,9,
10, etc…

Today, The ARABIC NUMBER is the World Most Widely Known, and Use, WRITTEN, and COUNTING Number, ThroughOut all Our Nation in, Homes, Churches, Schools, Businesses and play.

With this Simple Piece of Knowledge and Wisdom, let us:

Begin OUR FOUNDATION PRE-SCHOOL KINDERGARTEN BASE NUMBER COUNT Book,

Counting Beginning

THE ROMANS

WERE ONCE

IN

GREAT

POWER

NOW THE ROMAN NUMBER IS KNOWN WITH THE POPULARITY (FAME) of:

The SUPER BOWL.

ROME is also ONE OF OUR Known HISTORICAL HISTORY teaching:

THE ROMAN EMPIRE

And

THE FALL OF ROME

It has no starting point,
Nor ending point. It is
ROUND just like the
EARTH.

THEREFORE IT IS THE ONLY NUMBER THAT IS PERFECT. MEANING, IT IS THE ONLY NUMBER IN A NUMBER COUNT THAT IS A COMPLETE NUMBER. IT IS A WHOLE NUMBER. IT HAS NO BEGINNING POINT, AND IT HAS NO ENDING COUNTING:

INTRODUCING

THE NUMBER

0

HELLO EVERYBODY; MY NAME IS

0(ZERO)

and I AM THE NUMBER, I AM NOT THE ALPHABET O. I AM THE

ARABIC NUMBER IN THE ARABIC NUMBER COUNTING NUMBER, WRITTEN and SPOKEN, I AM THE FIRST NUMBER IN THE

ARABIC NUMBER COUNT COUNTING.

Before me there is no other number that can BEGIN A PERFECT ARABIC NUMBER COUNT COUNTING,

**I AM THE 1ST ARABIC NUMBER IN ITS FOUNDATION, BASE NUMBER COUNTING COUNT.
I AM THE NUMBER >>>>>**

OH BOY!
WASN'T THOSE
NUMBER SILENT,
BUT,
DIDN'T THEY
SPEAKLOUD

LOOK HERE>>>>>>

WHENEVER THE NUMBER ZERO

IS PLACE WITH

ANOTHER NUMBER,

THE NUMBER

BECOMES

LARGER AND

GREATER THAN (>)

THE

PRESENT NUMBER.

I AM THE 1ST NUMBER

BEGINNING NUMBER A

POWER TWO

EXAMPLE: THE NUMBER 0 WHEN PLACED WITH THE NUMBER 1, DIRECTLY AFTER THE NUMBER 1, INCREASES THE NUMBER 1 TO A NUMBER 10. 10 IS A NUMBER **GREATER THAN (>)** 9,8,7,6,5,4,3,2,1 AND 0 ALONE

THE NUMBER

10

TO THE

INFINITIES

0. 0 to 9 ZERO

1. 10—99 TEN

2. 100-999 HUNDRED

3. 1,000-9,999 THOUSAND

4. 10,000-99,999 TEN THOUSAND

5. 100,000-999,999 HUNDRED THOUSAND

6. 1,000,000-9,999,999MILLIONS

7. **1,000,000,000-9,999,999,999 BILLION**

8. **1,000,000,000,000-9,999,999,999,999 TRILLION**

9. **1,000,000,000,000,000-9,000,000,000,000,000 ZILLION**

IS

SINGLE

NUMBER.

ADDING **00** AFTER THE NUMBER **1** INCREATES THE NUMBER **1** TO THE NUMBER **100.** THE NUMBER **100** IS A NUMBER GREATER THAN(>) THE NUMBER **99.**

-0
MEANS
NOTHING.

.-0
MEANS NOTHING
BECAUSE
THERE IS NO NUMBER,
NOR NUMBERS
BEFORE
THE NUMBER ZERO
IN
A
PERFECT NUMBER COUNT

STOP!

STOP HERE!

I CLARA D. CREDIT the Author of FOUNDATION PRE-SCHOOL KINDERGARTEN BASE NUMBER COUNT BOOK Suggest that Pre-Schoolers and Kindergartens are Allowed to STOP Here; Unless you as a Parent, Instructor, or tutor see otherwise.

FOUNDATION PRE-SCHOOL KINDERGARTEN BASE NUMBER COUNT ELEMENTARY SCHOOL TO ADULT

RULE

THE

RULE

RULER

GREAT ENGINEERS USES THE RULE RULER TO BUILD MANY DIFFERENT THINGS.

DOCTOR'S AND PHARMACIS USES THE NUMBER OF MEASUREMENT OF COUNT TO WEIGHT.

AGE

SOCIETY USES THE COUNT OF 0, 1, 2, 3, 4, 5, 6, 7, 8, 9, 10 ETC…FOR AGE.

PREGNANT IS AGE 0. AGE 0 OF A WOMAN'S PREGNANCY IS LIFE.

After a baby is born outside of Its' Mother womb,
Its' 1st birthday then becomes the Number 1 year old.

Hey!
Lets'
COUNT THE
PRESIDENTS
OF
THE
UNITED
STATES
AMERICA

YOU CANNOT BEGIN THE PERFECT NUMBER COUNT

WITHOUT ME!

I MUST BE ACKNOWLEDGE.

TO BE SURE

MS. C

PRESIDENT

NUMBER **0.** **GEORGE W. BUSH**

PRESIDENT

NUMBER **1.** **GEORGE WASHINGTON**

PRESIDENT

NUMBER **2. JOHN ADAMS**

MS. C

PRESIDENT

NUMBER 3. THOMAS JEFFERSON

PRESIDENT

NUMBER **4.** JAMES MADISON

PRESIDENT

NUMBER **5. JAMES MONROE**

PRESIDENT

NUMBER 6. JOHN QUINCY ADAM

PRESIDENT

NUMBER 7. ANDREW JACKSON

PRESIDENT

NUMBER **8.** MARTIN VAN BUREN

PRESIDENT

NUMBER **9.** **WILLIAMS H. HARRISON**

PRESIDENT

NUMBER **10. JOHN TYLER**

PRESIDENT

NUMBER **11.** **JAMES K. POLK**

PRESIDENT

NUMBER 12. ZACHARY TAYLOR

PRESIDENT

NUMBER **13.** **MILLARD FILLMORE**

PRESIDENT

NUMBER **14.** **FRANK PIERCE**

PRESIDENT

NUMBER **15.** **JAMES BUCHANAN**

PRESIDENT

NUMBER **16.** **ABRAHAM LINCOLN**

PRESIDENT

PRECENTAGE

0%

IS A PRECENTAGE
OF
INTREST

LETS' CONTINUE
TO COUNT

THE

PRESIDENTS

OF

THE

UNITES STATES

OF

AMERICA

NUMBER **17.** ANDREW JACKSON

PRESIDENT

NUMBER **18.** ULYSSES S. GRANT

PRESIDENT

NUMBER **19.** RUTHERFORD B. HAYES

PRESIDENT

NUMBER **20.** **JAMES A. GARFIELD**

PRESIDENT

NUMBER **21.** **CHESTER A. ARTHUR**

PRESIDENT

NUMBER 22. GROVER CLEVELAND

PRESIDENT

NUMBER 23. BENJSMIN HARRISON

PRESIDENT

NUMBER 24. BENJSMIN HARRISON

PRESIDENT

NUMBER **25.** WILLIAMS MCKINLEY

PRESIDENT

NUMBER **26.** THEODORE ROOSEVELT

PRESIDENT

NUMBER **27.** **WILLIAMS H. TAFT**

PRESIDENT

NUMBER **28.** WOODROW WILSON

PRESIDENT

NUMBER **29.** WARREN G. HARDIN

PRESIDENT

NUMBER **30.** CALVIN COOLIDGE

PRESIDENT

NUMBER **31.** HEBERT HOOVER

PRESIDENT

NUMBER **32.** FRANKLIN D. ROOSEVELT

PRESIDENT

NUMBER **33.** HENRY S. TRUMAN

PRESIDENT

NUMBER **34.** DWIGHT D. EISENHOWER

PRESIDENT

NUMBER **35.** **JOHN F. KENNEDY**

PRESIDENT

NUMBER 36. LYNDON B. JOHNSON

PRESIDENT

NUMBER 37. RICHARD M. NIXON

PRESIDENT

NUMBER **38.** GERALD R. FORD

PRESIDENT

NUMBER 39. JAMES E. CARTER

PRESIDENT

NUMBER **40.** RONALD REAGAN

PRESIDENT

NUMBER **41.** **GEORGE BUSH**

PRESIDENT

ONCE UPON A TIME A LONG long time AGO IN THE LAND of ROME.

A GREAT MONACH NAME JULIA CEASAR RULED ROME.

THE ROMAN EMPIRE:

WE HOLDS THESE TRUTHS, TO BE SELF EVIDENCE, THAT ALL MEN ARE CREATED EQUAL.

MONEY COUNTING

**BEGINNING
MONEY COUNT
WITH
THE NUMBER
ZERO (0)
LEAVES
THE
CHECKBOOK
BALANCE
A
FIGURE
OVER**

CHECKBOOK BALANCING

Federal Reserve Notes

0

1

2

3

4

5

6

7

8

9

WOW!
I COUNTED
9
ONE
HUNDRED
DOLLARS
BILLS

AND NOW LADIES AND GENTLEMENS

THE
42ND PRESIDENT
OF
THE UNITED STATES
OF
AMERICA.

THE
ENDING
PRESIDENT
OF
THE
OUR CENTURY

PRESIDENT

NUMBER 42. BILL W. CLINTON

"LADIES AND GENTLEMEN!

FINAL MY SPEECH

EDUCATION

WHEN, HAVE WE EVER, IN THE ENTIRE HISTORY, OF OUR HISTORICAL CENTURY, OF MANKIND, NOT PLACED THE VALUE OF EDUCATION, AT THE TOP OF, THE UNITED STATES OF AMERICAN FLAG, SYMBOLIZING OUR COLOR (RED).

RED, REPRESENTING, WAR.
THE COLOR RED.
WAR, REPRESENTING THE SHEDDING OF INNOCENT
BLOOD: THE BLOOD SHEDDED OF INNOCENT MANKIND FOR US, FOR OUR CHILDREN, AND FOR OUR CHILDREN CHILDRENS' CHILDREN AND FOR TIMES AND TIMES TO COME.

AND WILL WE (NOT FOR A LACK OF KNOWLEDGE) MISTAKENLY ALLOW OURSELVES, TO ENGRAVE IN THE MINDS OF THE FUTURE OF THE UNITED STATES OF AMERICAN AND OUR NEIGHBORS ABROAD; THE CONTINUING TEACHING OF OUR EDUCATIONAL SYSTEMS CALLING THE MATHEMATIC NUMBER ZERO (0), AN ALPHABET (O), OR THE ENGLISH WORD, NOTHING, OR OWE AS IN A DEBT.

**AND SOME EVEN DARE SAY,
THAT THE NUMBER 0 (ZERO)
DOES NOT MEAN ANYTHING.
WHEN IN ACTUALITY**

NOTHING
IS
A
WORD

WORDS
ARE
ENGLISH
TOOLS

O
IS
AN
ALPHABET

AND

0
IS
A
NUMBER

NUMBERS
ARE
COUNTING
TOOLS

WHENEVER AN
ENGLISH WORD
ZERO
AND
A
COUNTING NUMBER
0
ARE
SPOKEN,
THEY
SOUND
THE
SAME,
ARE
PRONOUNCED
THE
SAME,
MEANS
THE SAME
BUT,
THEY
ARE
WRITTEN
DIFFERENT

WHENEVER
WE
WRITE
THE
ALPHABET
O
AND
WRITE
THE
COUNTING
NUMBER

0

THEY LOOK THE SAME
IN THE FORM OF THEIR SHAPE.
THEY ARE BOTH ROUND,
WITHOUT
A
BEGINNING POINT
AND WITHOUT
AN ENDING POINT.

GOD HAS NO BEGINNING, NOR, NO ENDING. GOD IS PERFECT. GOD IS COMPLETE. GOD IS THE WHOLE. JUST AS WE REFER TO, THE GREAT GOD, THE GREAT GOD OF THE UNITED STATES OF AMERICAN AND THE WHOLE WIDE WORLD; EXISTENCE—WITHOUT BEGINNING AND WITHOUT ENDING

**THE NUMBER 0
AND
THE ALPHABET O
HAVE
NOT MODIFY
INFLECTIONAL FORM.
AND BECAUSE THEY ARE
PERFECT COMPLETE
WITHOUT BEGINNING
AND WITHOUT AN
ENDING), THEY BOTH
ABSENT LACKING.**

IT HAS NO BEGINNING POINT, AND IT HAS NO ENDING COUNTING

0(ZERO)

THE PERFECT NUMBER COUNT!"

WOW!

NOW WASN'T THAT THE MOST EDUCATING, AND FUNNEST, AND EXCITING WAY EVER TO LEARN PERFECT FOUNDATION BASE NUMBERS COUNTING.

IF GOD WILL, WE SHALL LIVE, YOU'LL BE SEEING ME AMICABLE, YOUR FRIEND, AND FRIENDLY EDUCATION VOICE GUIDE,
COMING SOON IN
MY A   B(BEE)   C(SEA)   D(THE)   E   F   G BOOK
BYE BYE NOW, GOOD NIGHT!

THE FIRST BOOK HAVE I MADE,
O FRIEND OF **GOD.**

ABOUT THE AUTHOR

MS. C is the author of Foundation Preschool Kindergarten Base Number Count Book, 1stBooks Library, Bloomington Indiana USA. A songwriter on the 1999 Turn of the Century Collective Album.

The Light of the Word, Hilltop Records Hollywood California. She is an educator. She was born in the United States and holds a degree in office administration. She is a continued bible student.